IT'S SCIENCE!

All kinds of habitats

Sally Hewitt

W
FRANKLIN WATTS
LONDON•SYDNEY

First published in 1998 by Franklin Watts
Paperback edition 2001

Franklin Watts
96 Leonard Street
London EC2A 4XD

Franklin Watts Australia
56 O'Riordan Street
Alexandria, Sydney
NSW 2015

Series editor: Rachel Cooke
Designer: Mo Choy
Picture research: Sue Mennel & Alex Young
Series consultant: Sally Nanknivell-Aston

ISBN 0 7496 3069 8 (hbk.); 0 7496 4271 8 (pbk.)

Dewey Decimal Classification Number 574.5

A CIP catalogue record for this book is available from the
British Library.

Printed in Malaysia

Acknowledgements:

Ardea pp. 7t (Chris Martin Bahr), 13r (Jean-Paul Ferrero), 19t (S. Meyers), 27t (Ian Beames); Biofotos pp. 7bl (Ian Took); Bruce Coleman pp. 8tr (Hans Reinhard), 9tm (Stephen J. Krasemann), 10l (John Shaw), 11r (Leonard Lee Rue), 14b (Harald Lange), 17tr (Rod Williams), 18t & b (Johnny Johnson), 19b (Gordon Langsbury), 20tl (Konrad Woth), 20tr (Adrian Davies), 20b (Hans Reinhard), 21l (Rod Williams), 23t (Joe Macdonald), 23b (Gordon Langsbury), 26l (C.C. Lockwood), 27bl (Jane Burton); James Davis Travel Photography p. 12l; Ecoscene pp. 9b (Andrew Brown), 9tr (Chinch Gryniewicz), 12r (Andrew Brown); Frank Lane Picture Agency pp. 6 (Derek Hall), 8tl (David Hosking), 8b (Michael Callan); Getty Images p. 26r; Natural History Photographic Agency pp. 7br (G. Bernard), 16l (David Middleton), 17tl (Haroldo Palo Jr.), 21r (John Shaw), 27br (E.A. Janes); Oxford Scientific Films pp. 9tl (Irvine Cushing), 13l (Daniel J. Cox), 14t (Richard Hermann), 25t; Planet Earth Pictures pp. 10r (Jonathan Scott), 15 (John Lythgoe), 17b (Enrico Ferorelli), 22l (Richard Cottle), 22m (J & G Lythgoe), 22r (Gary Bell), 24t (Peter David); Still Pictures Cover (Regis Cavignaux), pp. 3 (Regis Cavignaux), 11l (J. Dennis), 16r (Regis Cavignaux), 24b (Fred Bavendam), 25m (Mark Carwardine), 25b (Rafael Al Ma'ary). All other photography by Ray Moller.
Thanks to our model, Zara Bilgrami.

IT'S SCIENCE!

All kinds of habitats

Contents

What is a habitat?

A habitat is the place where animals and plants live and grow.
You live in a habitat, too. A habitat is not just your home,
but the area your home is in. This busy city is a kind of habitat.
Lots of people have their homes there.

Can you see the plants growing in the streets? The city is their habitat as well.

Apart from people, what other animals might live in this city?

 THINK ABOUT IT!

Think about the habitat you live in. Do you live in the city, in the
countryside or near the sea? Do other people live near you?
Do any plants grow nearby? Are there any animals?
What kind of weather do you have?

Cities make one kind of habitat but there are lots of other places where plants and animals live. Forests and grasslands are habitats, so too are mountains, **deserts** and the frozen lands around the **Poles**.

Water forms lots of different habitats, too. There are **fresh water** habitats in rivers and lakes and salty water habitats found in the sea.

Animals and plants are **adapted** to live in their habitats. Fish can only live in habitats formed by water.

A habitat can be enormous, such as a great forest, or it can be tiny. Even a small stone forms a habitat for the animals living underneath it.

We will look at different habitats and the plants and animals that live in them in this book.

Woods and forests

Lots of animals and smaller plants live in the habitats formed by trees. Different trees grow in different places.

Coniferous trees grow where it is very cold. They have tough leaves that look like needles, and cones with seeds inside them. The trees grow close together in great forests. They can survive long, snowy winters.

Little grows on the dark forest floor. A carpet of pine needles is home to many insects. Birds, like this crested tit, feed on the insects.

THINK ABOUT IT!

Sunflowers grow in warm sunlight. Why don't they grow in coniferous forests?

Deciduous trees grow in warmer places. They have bigger leaves that die and fall to the ground in the autumn.

Woodpeckers find plenty of food in deciduous woods. They peck insects out of tree bark.

Deer eat plants and leaves from the trees, too.

Flowers such as bluebells grow in the spring when the sunlight shines on the woodland floor.

TRY IT OUT!

A single tree is a habitat. Study a tree near you. Find out what kind of tree it is. Watch out for insects, birds and other animals that live in or visit your tree. Do any plants grow on it?

Grasslands

In some places in the world, the land is covered in grass for thousands and thousands of kilometres. There are few trees, but the grass provides food for many animals.

In the grasslands of Africa, many different animals eat grass. They can all live together because they eat different parts of the grass.

Stripy zebra eat the top of the grass. The antelope graze on the short grass near to the ground.

The animals move across the plains so that there is always fresh grass to eat. The Masai people keep cattle on the grasslands. They move their herds around, too, to find fresh grass.

The grazing animals are food for meat eaters, such as a lion. The big cat can hide in the grass when he hunts.

The lion is not always successful in his hunting. The grass eaters he chases can run very fast!

In grasslands, there are few trees and bushes for homes and shelter. Small animals, such as these meerkats, dig burrows underground. They can dive into them for safety when an enemy is near.

 **THINK ABOUT IT!**

What other animals can you think of that live in burrows?
What sorts of habitats do they live in?

Mountains

Plants and animals live high up on rocky mountain slopes, where it can be very cold. They have different ways of surviving the fierce winds, ice and snow.

High mountain peaks are often snow covered all year round. Most plants, animals – and people – live lower down the mountains' slopes.

Plants grow close to the ground and between cracks in the rocks to shelter from the weather.

👁 LOOK AGAIN

Look again at page 8 to see what kind of trees might grow on the lower mountain slopes where the weather can still be very cold.

THINK ABOUT IT!

What do you do to shelter and keep warm in cold and windy weather?

Snow leopards are **camouflaged**. They have pale fur so that they are difficult to spot among the rocks and snow. Their coats are thick and warm and they have fur on their paws for walking on snow and ice.

Mountain goats have thick coats, too. They have very good balance for climbing and leaping from rock to rock.

TRY IT OUT!

Experiment with camouflage. Cut out an animal shape and choose a background, for example, green leaves, rocks or sand. Paint your animal so that it is difficult to see against the background.

13

Deserts

Deserts are very dry places where hardly any rain falls. They can be very hot in the day and very cold at night. The ground is often rocky or sandy.

The plants and animals that live in deserts have to survive with very little water. Camels go for a long time without water but drink enormous amounts when they find it. Their fatty humps store food.

Cactus plants are well adapted for growing in the desert. They store water in their thick, fleshy stems. They often have sharp spines to stop thirsty animals eating them.

Do you get very thirsty in the hot sun? Unlike camels, we need to drink water every day. Desert people live near springs of fresh water called **oases** that bubble up from underground. Long, loose robes help them to keep cool and protect them from the burning sun.

TRY IT OUT!

Many desert animals burrow underground or hide in the shade in the hot day and come out at night when it is cooler. Put a little water in two saucers. Leave one in a cool, shady place and put one out in the sun or another hot place. Which saucer of water dries up first? Why do you think people and animals try to keep out of the sun in the desert?

Rainforests

Hot wet **rainforests** are the habitat of millions of different plants and animals. Plants grow here that cannot be found anywhere else in the world.

In a rainforest, huge trees form a thick leafy roof called the **canopy**. This blocks out the sun from the forest floor below.

Orchids, vines, bromeliads and air plants grow high up in the branches to reach the sunlight. Their roots cannot grow down into the earth so they collect water from the damp air.

Brightly coloured frogs live among the plants and in the water that collects in their leaves and flowers.

Animals live in every part of the rainforest.

Toucans search for fruit and insects in the canopy which they eat with their colourful beaks.

Spider monkeys use their tails like an extra arm to swing through the branches that reach above the canopy.

 THINK ABOUT IT!

Many animals and plants can only live in one kind of habitat. If we spoil their habitat they have nowhere to live. Without somewhere to live, they may die out and become **extinct**.

Jaguars hunt in the shadows of the forest floor.

Polar lands

In polar lands near the North and South Poles it is cold all year round. The winters are long and very dark. The summers are short and, even though the sun shines, it is still very cold.

Emperor penguins live near the South Pole. Their feathers help to keep them warm. They live in groups and huddle together for warmth.

Polar bears live around the frozen North Pole. They have thick fur and an extra layer of fat to keep them warm.

👁 LOOK AGAIN

Look again at page 13 to find another animal with a thick coat that lives in the snow.

Summer in the Arctic near the North Pole is a very busy time. It only lasts for about two months when the sun never sets.

The ice and snow melt and millions of insects buzz around the flowers that spring up.

Flocks of birds and herds of caribu make long journeys north to feed on the plants and insects. When the cold weather begins, they travel south again where it is warmer. These long journeys are called **migrations**.

It is winter at the South Pole during the North Pole's summer. Arctic terns fly all the way from the North Pole to the South Pole so that they can spend the summer in both places!

Fresh water

Water in lakes and rivers is **fresh**, not salty like seawater. The banks, the water and the muddy beds of rivers and lakes make different habitats for all kinds of animals and plants.

The water is full of animals and plants that can only be seen through a microscope.

These tiny living things are food for other water **inhabitants**, such as snails, worms and insect larvae that live in the mud at the bottom of rivers and lakes.

Fish swim among the underwater plants, feeding on the plants and other water creatures.

 **THINK ABOUT IT!**

Factories and farms sometimes spill water into lakes and rivers. What do you think could happen to the plants and animals that live there?

20

Moorhens feed on water plants, insects and tadpoles. They build their nests in the shelter of reeds and rushes that grow on the banks.

Otters dig their burrows in muddy banks and swim in the water catching fish and frogs to eat.

TRY IT OUT!

Is there a pond near your home or school? If so, visit it with an adult and spend some time looking at the plants and animals that have made their home there. Watch out for birds, insects and other animals who visit it.

Seashore

The seashore is the place where the sea meets the land. It is a habitat that changes all the time as the waves crash and the **tide** comes in and out. Like fresh water, the sea is full of tiny animals and plants to eat.

Plants and animals cling tightly to the rocks. Seaweed attaches itself to the rocks by **holdfasts**.

 THINK ABOUT IT!

What might happen to seaweed, anemones and limpets if they did not cling to the rocks when the tide came in and out?

Anemones are animals fixed to the rocks by their stalks. They wave their tentacles underwater to catch food when the tide is in and fold them away at low tide.

Crabs live underwater and on the shore.
They bury themselves under the sand
when there is any danger. Sometimes you
can just see their eyes keeping watch.

There is plenty of food on the seashore
for hungry seabirds. The sea is full of fish,
worms wriggle in the sand and there are
crabs and shell fish among the rocks.
The birds often have long legs and long
beaks for finding food under the sand.

LOOK AGAIN

Look again at page 18 to
find another seashore that
looks very different from
the ones on these pages.

Under the sea

Did you know that there is more sea on earth than there is land? The sea is the biggest habitat of all. Different kinds of fish and sea creatures live in sunny shallow water from those that live in the cold and dark at the bottom of the sea.

A **coral reef** in sunny, shallow water is always busy. Corals look like plants but they are really the skeletons of millions of tiny sea creatures. Colourful fish dart around finding plenty to eat.

Deep under the ocean it is very dark. Without sunlight, plants cannot grow there but there are strange fish with glowing spots of light and others with enormous eyes.

👁 **LOOK AGAIN**

Look again at page 22 to find a sea creature that looks like a plant.

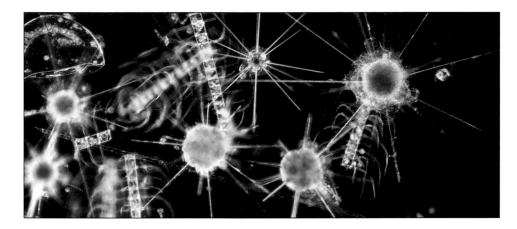

Open seas stretch for thousands of miles. Tiny plants and animals called **plankton** float in the water.

Plankton is the food of many animals, including huge whales. The hair-like fringe around the mouth of this humpback whale is called baleen. It sieves the plankton and other small sea creatures out of the water.

Fish that live in the open water swim together in shoals. They have streamlined bodies to swim fast through the water.

In the city

People live in most of the different habitats we have looked at in this book. The houses they build and the clothes they wear often depend on the habitat they live in.

People have changed natural habitats by building cities with houses, factories and roads to live and work in. Can you see how this city has changed the natural habitat?

Many plants and animals have found ways of living in the city. In North American cities, racoons live among the houses, foraging for food.

 THINK ABOUT IT!

Look all around you on a busy street. Can you see at least one plant, animal, bird or insect?

Mice make their nests in warm places under the floorboards of houses. They find crumbs to steal and eat from food cupboards.

Foxes have learnt how to knock the lids off rubbish bins to find something to eat inside.

When people leave a building empty, plants soon find all kinds of places where they can grow.

👁 LOOK AGAIN

Look again through this book. How do the habitats people live in affect their everday lives? How have people changed their habitats?

Useful words

Adapted Animals are adapted to live in their habitats. This means they have special things about them that let them live there. For example, polar bears are adapted to live in icy lands with their thick, white fur.

Camouflage Animals are camouflaged when their coats or skins blend in with the colours of the habitat they live in so that they are hard to see.

Canopy The leaves of tall rainforest trees form a thick roof over the forest called a canopy.

Coniferous trees Coniferous trees have needle-like leaves and cones. They have leaves all year round. Pine trees are coniferous trees.

Coral reef Coral reefs are found in warm, shallow seas. They look like plants but are made by the skeletons of millions of tiny sea creatures, which build up over a very long time.

Deciduous trees Deciduous trees have leaves that die and fall to the ground in autumn. New leaves grow again in the spring. Deciduous trees have flowers rather than cones.

Deserts Deserts are dry places where very little rain falls. Some are very hot and others are cold. The ground is often rocky or sandy.

Extinct When a type of animal or plant has completely died out and there are none left on Earth, we say they are extinct.

Fresh water Rain water and water in rivers and lakes is fresh water. Unlike sea water, it is not salty and people and animals can drink it.

Holdfasts Seaweed does not have roots. Instead it clings to rocks and the seabed with holdfasts.

Inhabitants Inhabitants are the people or animals that live in, or inhabit, a particular place.

Migrations Migrations are the long seasonal journeys that some animals make to find food. They migrate from somewhere cold to somewhere warmer during the winter.

Oases Oases are wells of fresh water that spring up from underground in a desert. An oasis is a single desert spring.

Plankton Plankton is the name for the millions of tiny plants and animals that float in the sea.

Poles The North Pole is the northern most point on Earth. The South Pole is the southern most point. Around the Poles it is always very cold, with long dark winters and short cool summers.

Rainforest Rainforests grow in hot places where it rains nearly every day. They are home to huge numbers of animals and plants.

Tide Each day the sea moves in and out of the seashore. We call this the tide. When the tide is low, there is more land along the shore. When the tide is high, more of the seashore is underwater.

Index

About this book

Children are natural scientists. They learn by touching and feeling, noticing, asking questions and trying things out for themselves. The books in the *It's Science!* series are designed for the way children learn. Familiar objects are used as starting points for further learning. *All kinds of habitats* starts with a city street and explores the different places plants and animals live.

Each double page spread introduces a new topic, such as rainforests. Information is given, questions asked and activities suggested that encourage children to make discoveries and develop new ideas for themselves.
Look out for these panels throughout the book:

TRY IT OUT! indicates a simple activity, using safe materials, that proves or explores a point.
THINK ABOUT IT! indicates a question inspired by the information on the page but which points the reader to areas not covered by the book.
LOOK AGAIN introduces a cross-referencing activity which links themes and facts through the book.

Encourage children not to take the familiar world for granted. Point things out, ask questions and enjoy making scientific discoveries together.